Energy Basics

Energized!

Glen Phelan

Program Developer, Kate Boehm Jerome
Program Design, Steve Curtis Design Inc.
www.SCDchicago.com

Sally Ride Science
9191 Towne Centre Drive
Suite L101
San Diego, CA 92122

ISBN: 978-1-933798-57-8

Printed in the United States of America
10 9 8 7 6 5 4 3 2 1
First Edition

Cover: Power lines carry electrical energy to towns and cities. Energy can move from place to place or change form, but it is never lost.

Title page: A solar car built by a team at Western Michigan University taps energy from the Sun to cruise along during a race.

Right: A wind turbine converts the wind's energy of motion to electrical energy that lights up homes and powers appliances.

Contents

Introduction

In Your World

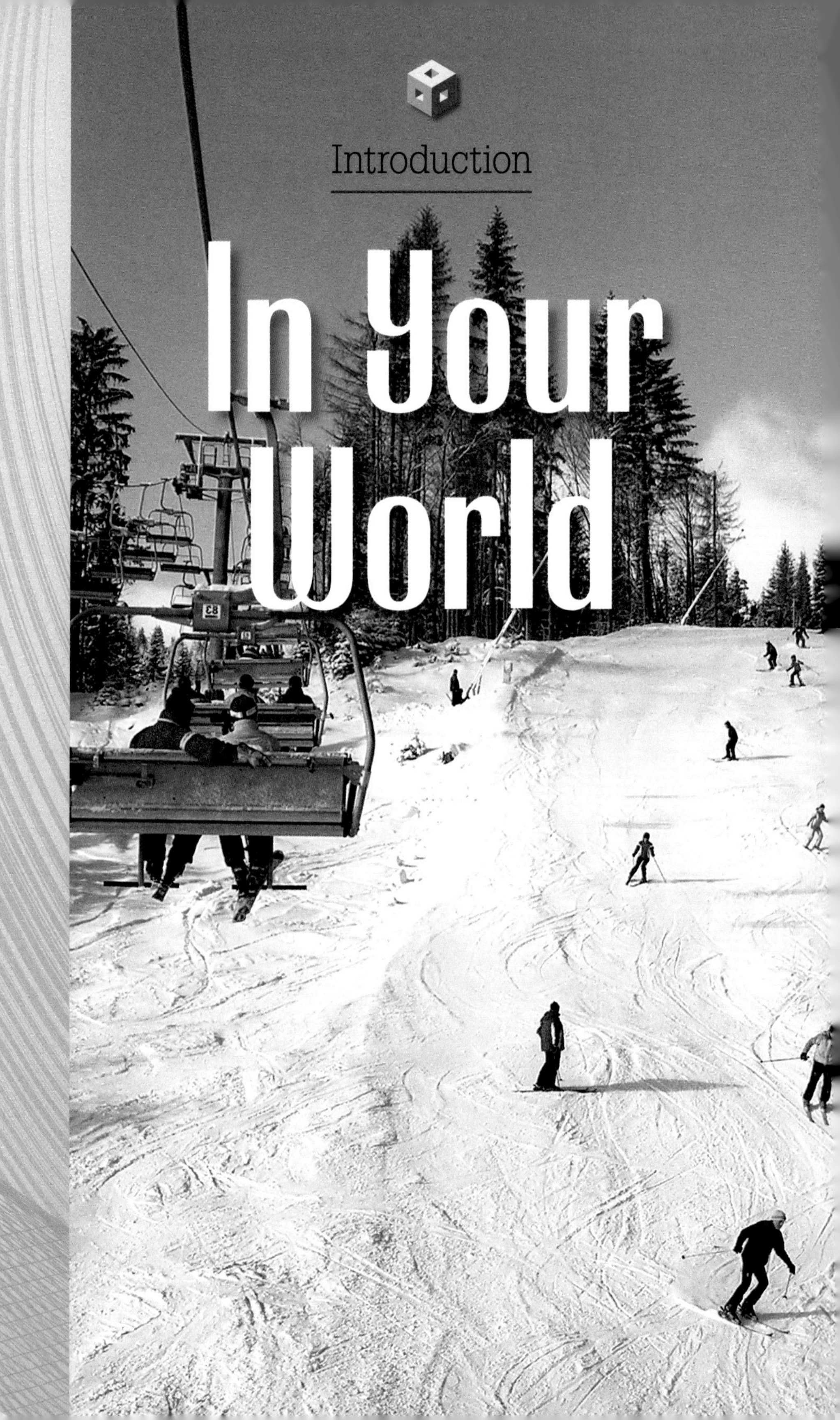

Looks like a fun-filled day on the slopes! Picture yourself there. What do you see? What do you hear, smell, and feel? All these things have something special in common.

How is the chairlift going up the mountain like the skiers gliding back down? What does the brisk breeze on your face have in common with the laughter of beginning skiers? And what does all this activity have to do with the sunshine reflecting off the sparkling snow?

Without this special something, skiers couldn't ski, the chairlift would stop cold, and the Sun wouldn't shine. This day of winter fun would be simply impossible.

Have you guessed what it is? **Energy**! That's right—it's all about energy. Everything in the photograph happens because of it. The scene shows many different kinds of energy. Can you tell what they are? You'll be able to by the time you finish this book. But first, let's start with a basic question—what exactly is energy?

Chapter 1: A World of Energy

Making Things Happen

"I don't have the energy to walk one more step."

"The team is playing with a lot of energy today."

"Close the refrigerator—you're wasting energy."

Have you ever heard people say similar things? The word *energy* pops up a lot, doesn't it? There's a reason for that—energy is a part of everything we do. Energy is a part of everything around us, too. What is energy? Energy is the ability to cause change. Energy is needed to make things happen—it's needed to cause change.

Take a look at this photo. What changes are happening? The dog started out on the ground, but now it is leaping high into the air. That's one change. The Frisbee was still until it was thrown. Then it sailed along until the dog grabbed it and stopped its flight. That's another change.

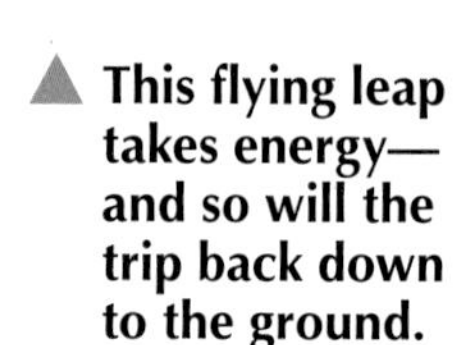

▲ This flying leap takes energy—and so will the trip back down to the ground.

Each of these changes takes energy. Someone used energy to throw the Frisbee, so the Frisbee has energy as it flies. The dog used energy to make the perfectly timed leap to snatch the Frisbee out of the air. And the dog will keep using energy as it lands and hurries back to its master for another Frisbee throw.

Energy on the Move

Take a look around you, indoors and out. Name a few things you see that have energy.

Chances are, if you saw something moving, you said it has energy. Well, you're right! There is energy in motion. If something is moving, it has energy, and that enables it to cause changes and make things happen.

What can energy of motion do? Plenty! The energy of a rolling bowling ball knocks down the pins. The energy of a strong wind blows away an umbrella. The energy of your moving arm opens a door. In each case, the energy of motion causes a change.

What changes result from the energy of this speeding train? If you've ever been near one, you know. The screaming, rumbling noise of tons of steel racing past you can be deafening. Besides that, the ground shakes like it might during an earthquake. These changes don't happen when the train stands still. They are caused by the energy of the train in motion.

Closet Commotion

The atoms and molecules that make up matter are constantly moving—jiggling, twisting, and turning. They're moving even in solids, like the shoes in your closet and the clothes in your drawers. So really, everything has some energy of motion.

▲ **This speeding train has the energy of motion—big-time!**

The Bottom Line

If something is moving, it has energy and can cause change.

Flying Balloons

An object can have energy even if it's not moving. Here's a fun way to show how. Blow up a balloon—okay, for now, just *think* about blowing up a balloon. Don't tie the neck—just pinch it closed.

Does the balloon have energy? Well, let's see—it isn't moving, so it doesn't have energy of motion. Also, the balloon isn't causing any changes. But remember, energy is the *ability* to cause change. Does the balloon have this ability? Of course it does! If you want to prove it, just let go.

The balloon sure has energy of motion now! It dips and darts all over the place. Plus, it makes that funny spitting sound. Where did the balloon get the energy to make these changes?

The energy came from you! You gave the balloon energy when you stretched it out by blowing air into it. The energy of the moving air became stored in the balloon.

▲ **Which drawing shows energy? Hint—it's a trick question.**

Store It Up

Things that are stretched, bent, or squeezed change shape and store energy. The objects release the stored energy as they return to their original shapes. This released energy can move more than balloons. Check out this boy pulling back a bowstring. How is energy stored in this case? What happens when the energy is released?

You store energy and then use it countless times throughout the day. Every time you eat, you store energy. Food is the fuel that gives you the energy to walk, talk, blink, and breathe.

Go With the Flow

Stored energy doesn't disappear as it's used. It just changes form and shows up elsewhere. Imagine diving into a pool. You use the energy stored in food to power your leg muscles and jump off the diving board. This energy of motion becomes stored energy at the top of the jump. It turns back into energy of motion as you fall. Then . . . *splash*! The energy shoots water upward and spreads out as **waves**.

Where does the energy come from to pull back a bowstring? And where does the energy go?

The Bottom Line | **Energy can be stored and released later.**

Chapter 2: Changes in Energy

Used, but Never Used Up

By now, one thing should be clear—nothing happens without energy. We need energy to get things done, whether that means spiking a volleyball, nailing together pieces of wood, or typing on a keyboard. These actions use the energy of motion, which is also called **kinetic energy**.

A volleyball player uses kinetic energy to leap, swing her arm, and strike the ball. Now the ball has kinetic energy—and lots of it. The ball is moving really fast as it rockets over the net. A player on the other team feels that energy when the ball hits her outstretched hands.

▲ **Without kinetic energy, volleyball wouldn't be much of a sport.**

A hammer has kinetic energy when a carpenter swings it again and again as he pounds a nail. That energy drives the nail into the wood.

When you type on a computer keyboard, your moving fingers have kinetic energy. That energy is transferred to the keys as you tap out the answers to your homework questions.

Blowing in the Wind

Which moves a sailboat faster—a strong wind or a gentle breeze? A strong wind, of course, but do you know why? It's because the fast-moving air of a strong wind has more energy than the slow-moving air of a breeze. The more energy there is, the more that energy can do—like push a sailboat faster through the water.

The amount of energy a moving object has depends on two things—its speed and its **mass**. Changing either one changes the amount of energy the object has and what that energy can do.

How does mass affect the energy of these dogs? They are moving at the same speed, but they don't have the same energy of motion. The bigger dogs have a lot more mass, or amount of matter, so they have more energy. Who cares? You would if you were walking the dogs and they suddenly lunged after a squirrel! You could pull back the little ones, but the big dogs' energy might drag you right along.

Which dog would be easiest to take for a walk? Why?

The Bottom Line

The more speed or mass a moving object has, the more energy it has and the more that energy can do.

Chemicals Make It Go

With a mighty roar, the rocket climbs higher and higher. It sure has a lot of energy of motion—the rocket is a lot of mass moving fast! But without **chemical energy**, the rocket couldn't even get off the ground.

Fuel in the rocket contains chemical energy. The way atoms in the fuel are held together enables the fuel to store energy. The energy is released during **chemical reactions**, like burning.

When the rocket fuel burns, its chemical energy doesn't disappear. It just changes into different forms. Can you tell what they are? The most important one is kinetic energy, which moves the rocket into space. The burning fuel also produces bright **light**, intense **heat**, and loud **sound**. These are other kinds of energy that come from the release of stored chemical energy.

▲ **What forms of energy are part of a rocket launch?**

Batteries Included

You don't have to wait for a rocket launch to see chemical energy at work. Any fuel contains chemical energy. When fuel burns, the released energy changes form. This energy can be used to do all kinds of things.

Speaking of fuel, have you had any today? Remember, food is fuel for the cells that make up your body. What did you eat for breakfast or lunch? The chemical energy in that food is already hard at work. Chemical reactions in your cells are releasing the energy to keep your heart beating and your brain working. This energy keeps the rest of your body in tip-top shape.

You're probably not far from another source of chemical energy—batteries. Chemical energy is stored in a paste inside a battery. Electricity flows when certain parts of the battery are connected. This energy powers watches, cell phones, cameras, and flashlights.

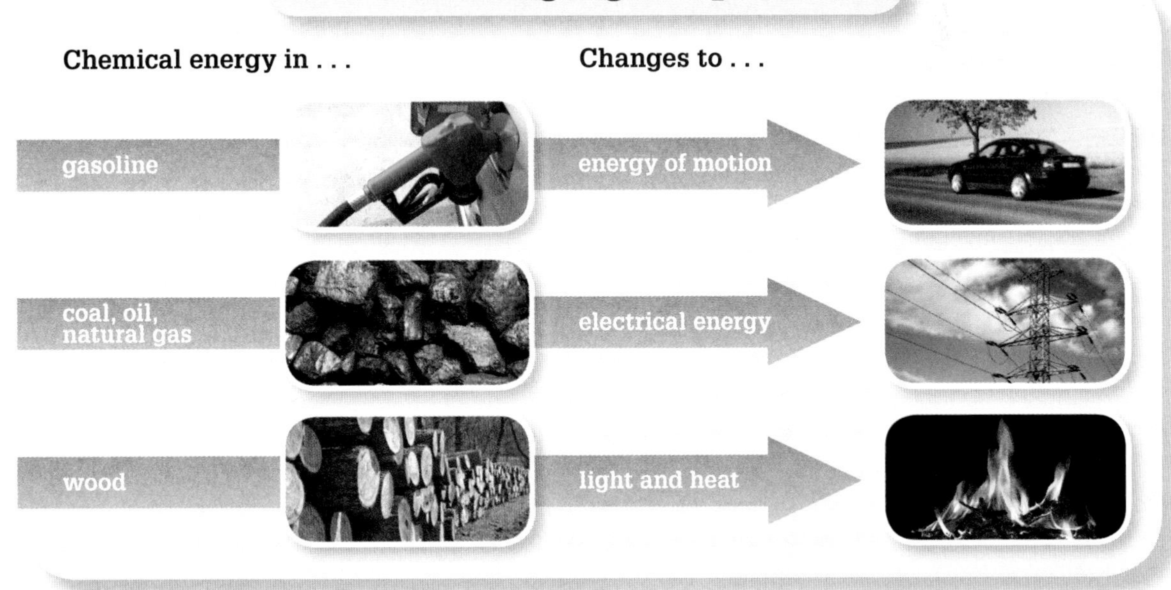

The Bottom Line | **Chemical energy in fuels, food, and batteries can change into other forms of energy and can be used to do many things.**

Heat moves from the burner flames to the pan to the egg as particles bump into each other. The result? *Mmm . . .*

Turn Up the Heat

Cup your hands over your mouth and breathe out. Do you feel the warmth? Where did it come from? It's heat—another kind of energy your body gets from food.

Heat comes from the motion of the atoms and molecules that make up all matter. These particles are constantly moving. The faster they move, the more heat the matter has and the warmer it feels.

You might say that things get cooking when particles start moving. For example, think about frying an egg. The particles that make up the pan are jiggling back and forth in place. The pan itself isn't **vibrating**, but its particles are. At first the pan feels cool because the particles are moving fairly slowly. But as you start to cook, heat from the burner makes the particles jiggle faster and faster. The pan gets hot. The vibrating particles in the pan crash into particles in the food and make them vibrate faster, too. In this way, energy flows as heat from the burner to the pan to the food.

A Shocking Experience

Did you ever get a shock as you reached out to grab a doorknob? You got zapped by the same kind of energy that lights your classroom, powers your computer, and runs most other machines you use. You experienced **electrical energy**.

Electrical energy, or electricity, is energy that comes from the flow of electrons. Your shock came from electrons jumping from you to the doorknob. Useful electricity, however, usually flows through wires. The flip of a switch or push of a button puts this form of energy to work. It's hard to beat that kind of convenience!

Watts and Bolts

Lightning packs a punch. So you might think harnessing this power would solve all our energy needs. Think again. A single bolt has only enough energy to power a 100-**watt** light bulb for 6 months. Even if we could harness this power, it would take about 58,000 lightning strikes each day to equal the energy output of one large power plant.

▼ **Providing light is just one of the many uses of electrical energy.**

The Bottom Line

Heat and electricity are two forms of energy that people use every day.

Pass It On

When you play catch with a friend, you're doing more than passing a ball back and forth. You're passing energy back and forth. Your friend gives the ball energy when he throws it. You feel that energy of motion when you catch it. But what if you throw the ball too far and accidentally hit the side of a parked car? *Uh-oh!* The energy is transferred to the car and makes a dent.

Energy is constantly moving from one place to another. Heat energy is transferred as it flows from warmer objects to cooler objects. That's how a fireplace warms a cool room, and that's how sunshine melts an ice cream cone.

Sound energy travels in waves. The source of a sound, such as a barking dog, makes the surrounding air vibrate. Those vibrations travel through the air as waves. When the waves reach your ears, they make your eardrums vibrate. That's how you hear the sound.

Wind causes most water waves. Waves carry the wind's kinetic energy across the sea until they crash on the shore.

▲ **What happens to the ball's kinetic energy?**

It's the Law

When you kick a soccer ball, it eventually rolls to a stop. It may seem like the ball's energy gets used up and disappears, but here's what really happens.

As the ball rolls, some of its energy goes into bending the grass. Each bent blade of grass stores this energy until the ball passes and the grass stands up again. Some of the ball's energy changes to heat as the ball collides with the grass and the air. The ball, grass, and air actually become a little bit warmer, though you won't notice the change. What you do notice, though, is the swishing sound of the ball against the grass.

So, the ball's energy doesn't disappear. It just changes into heat, sound, and the motion of the grass. In fact, energy cannot be created or destroyed. It can only change from one form to another. This idea has been proven time and again. It is called the **law of conservation of energy**. And unlike the speed limit, it's a law that can't be broken!

The Bottom Line | **Energy cannot be created or destroyed. It can only change from one form to another and move to another place.**

Chapter 3: The Sun

Energy to Spare

Magnification: 560x

Microscopic phytoplankton use energy from the Sun to make food. Krill, shrimp-like creatures the size of your little finger, eat the phytoplankton. A humpback whale as big as a school bus gobbles the krill.

With a mighty thrust, the humpback whale rises out of the water. That's a lot of mass to throw around. Where does the whale get the energy to move with such power? Believe it or not, it all comes from the Sun.

Just like you, the whale needs food to give it the energy to go about the business of life—diving, swimming, digesting, breathing, and so on.

This giant mammal gobbles up millions of tiny animals called krill. The krill get their energy by eating even tinier phytoplankton. These living things are like microscopic plants. They float near the ocean surface, where the sunlight reaches them. This is important, because phytoplankton use the energy in sunlight to make their own food.

So the Sun is the *real* source of the whale's energy. In fact, the Sun is the source of almost all energy on Earth. It's the source of your energy, too. But you need some help from plants to be able to use it.

A Bite of Sunshine

When sunlight shines on a plant or phytoplankton, something incredible happens. The energy in sunlight powers chemical reactions inside some of the plant's cells. The reactions combine carbon dioxide from the air and water from the soil to make molecules of sugar and oxygen. This process is called **photosynthesis**.

The oxygen escapes into the air and helps us all breathe a little easier. The plant uses the sugar molecules as food to live and grow. What the plant doesn't use, it stores in its roots, stems, fruits, or leaves. A carrot is the stored food in the root of a carrot plant.

The energy stored in plants becomes food for nearly all other living things, including people. When you chomp on an apple, you're eating food made by plants. And when you eat a hamburger, you're also eating food made by plants—you're eating the meat of animals that grazed on plants. In this way, the Sun's energy becomes your fuel.

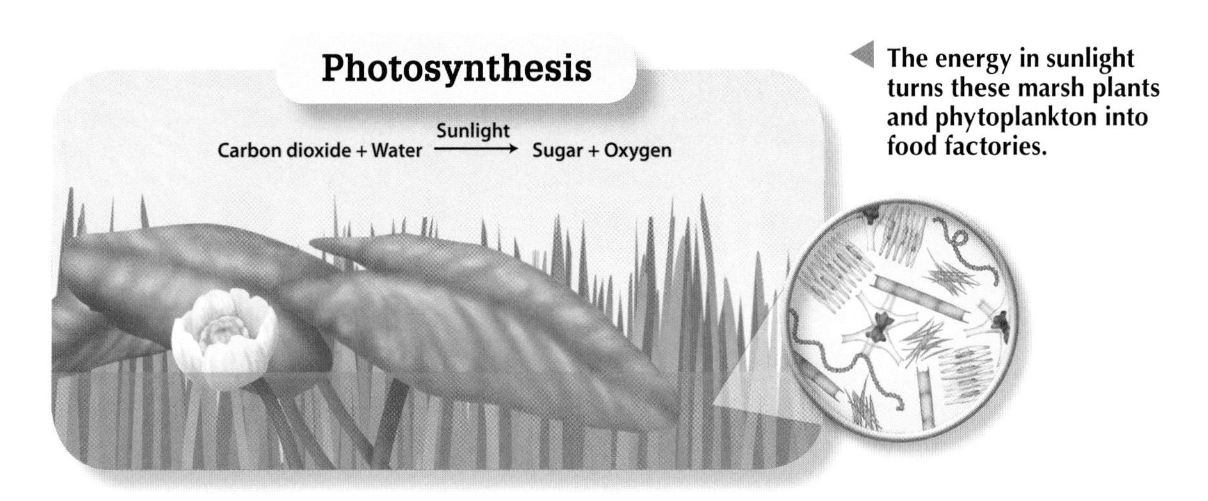

The energy in sunlight turns these marsh plants and phytoplankton into food factories.

The Bottom Line

The energy in sunlight is used by plants and phytoplankton to make the food that fuels our daily activities.

Keeping It Light and Cozy

Energy travels out from the Sun in all directions. This **solar energy** moves through space as waves. About 150 million kilometers (93 million miles) from the Sun, these waves happen to run into a place called Earth.

That's good for us. Not only does sunlight provide the energy needed for food to grow, it provides the light we need to see. We see only because light bounces off matter and reaches our eyes. In other words, no light, no sight.

The Sun also provides the heat that keeps our planet cozy, but maybe not in the way you think. Solar energy doesn't heat the air directly. Instead, it passes through the air and heats the surface. The land and water then **radiate** heat into the air. Some of the heat escapes into space. But most of it gets trapped by some of the gases in the air. This process, called the **greenhouse effect**, keeps the air, and us, toasty warm.

▲ **Most of the solar energy that reaches Earth's surface radiates back into the air as heat.**

The Wind Maker

What a view! This paraglider pilot is counting on using the Sun to extend the ride. How? By catching some thermals.

The Sun has a starring role in creating thermal currents that lift this paraglider.

Thermals are columns of rising air caused by the sunlight that heats the ground. A patch of ground, such as a clearing in a forest, may warm up more than the trees. The air above the clearing warms, becomes lighter, and rises, lifting the paraglider. Cooler air sweeps in to replace the rising air. This movement of air along the ground is wind.

The same thing happens on a global scale. Because Earth is shaped like a beach ball—a sphere—sunlight shines most directly near the equator. This part gets warmest. The warm ground heats the air, which rises. Cooler air rushes in, producing wind. The rising air travels away from the equator until the air cools enough to sink. These motions create circular patterns that carry air—and energy—all around the world. And it's all because of that great maker of winds—our Sun.

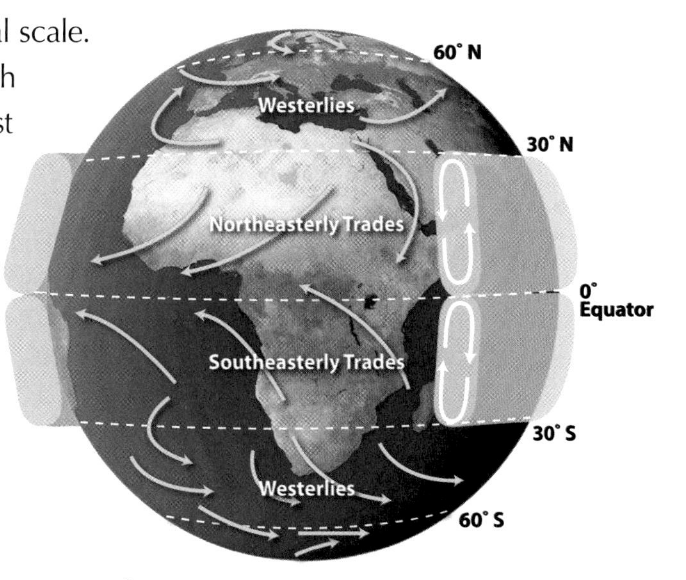

The Sun's uneven heating of the Earth drives global wind currents.

The Bottom Line | **The Sun's energy is the source of light and heat for Earth.**

The Rain Maker

If you were surprised to learn that the Sun causes wind, then stand by for a real shocker—the Sun also causes rain! How can the Sun *cause* rain if you can't even see the Sun when it rains? The answer is simple—the Sun's energy drives the **water cycle**.

Do you remember the puddles of water that formed the last time it rained? What happened to them? They dried up. The Sun's energy makes water **evaporate** from puddles, rivers, lakes, and oceans, and even from inside plants. Evaporation is one way water constantly moves between Earth's surface and the atmosphere.

When the water vapor rises and cools, it **condenses** into tiny droplets of water. The droplets collect and form clouds. As more water vapor condenses, the drops get larger. Eventually they fall as rain, sleet, hail, or snow.

Thank You, Sun

Flip back through all the examples of energy in this book. Almost every one can be traced back to the Sun. That yellow ball out there in space is pretty special. After reading this book, you may never think about the Sun in the same way again.

The Weird!

Dino Drink

Water travels from place to place, but it doesn't escape from Earth. Almost all of our water has been here for billions of years. Think of it. Atoms that make up the water you drink today may have been lapped up by a stegosaurus or pterodactyl drinking from a river 150 million years ago!

The Bottom Line

Energy from the Sun drives the weather and the water cycle.

THINKING LIKE A SCIENTIST

You're riding along the highway with your family on vacation. Suddenly you see a monstrous tower with giant twirling blades. There's another, and another . . . there are dozens of them! What are they, and why are they here?

Ordering

You probably know that these towers are wind **turbines**. They are here to turn the energy of the wind into a more useful form of energy—electricity. Some wind farms contain hundreds of turbines. Some are built in the ocean near the shore, where the wind really blows.

How do you get electricity out of the wind? It's all a matter of understanding how energy changes form. Here are the steps in order.

1. The kinetic energy of the wind—the moving air—turns the blades of the turbine.
2. That energy is transferred to a series of rods and gears and finally to a shaft that enters a generator.
3. The shaft spins coils of wire near powerful magnets. This produces a flow of electrons in the wire—electricity.
4. Wires carry the electricity to homes and businesses. There, people change the electrical energy into many other forms—light, heat, and sound.

◀ **A wind turbine's energy is transformed and moved over and over before it reaches homes and businesses.**

Interpreting Data

Today, the United States gets less than 2 percent of its electricity from wind power, but that number is rising and eventually will be a lot higher. Enough wind blows across the shores, mountains, deserts, and prairies to provide electricity for every home and business in the country. The trick is making this electricity cheap enough and getting it to all the places that need it.

The map shows the potential for wind energy in the United States.

Your turn! Use the information on this page to answer these questions.

1. Which places have the greatest potential for using wind energy?
2. Do you think your state would be a good place to build wind farms? Explain your answer.
3. If you were in charge of planning 10 large wind farms, where would you want to place them? Explain your answer.
4. Many new wind farms are being built offshore. What are some possible advantages of locating wind farms in lakes or oceans? What are some possible disadvantages?

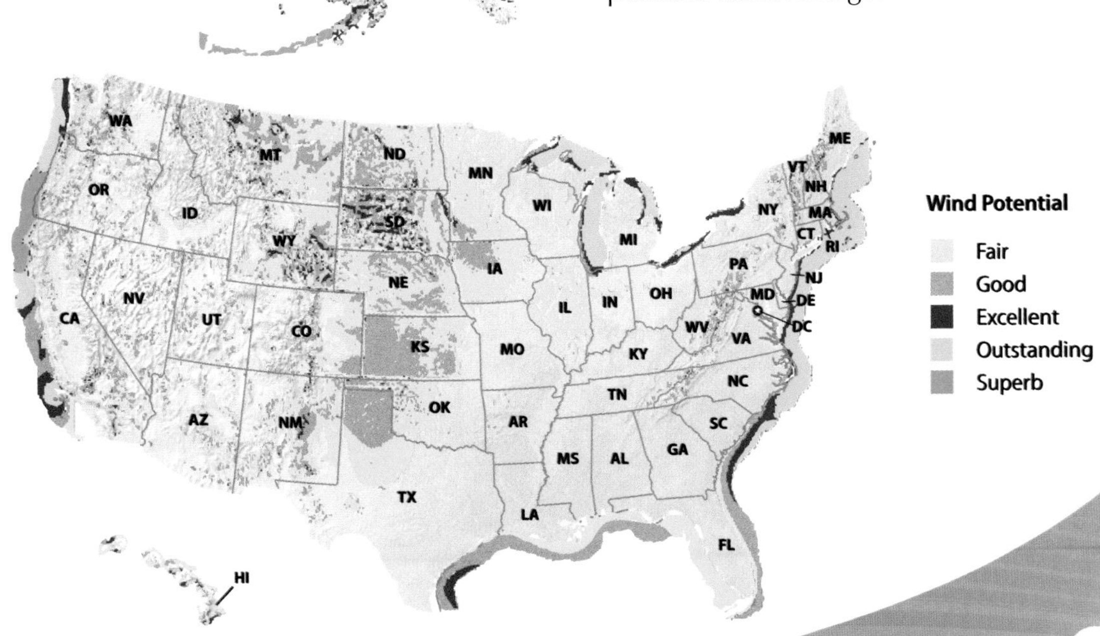

HOW DO WE KNOW?

THE ISSUE

An Energy Source Close to Home

Wires carrying electricity crisscross America in a giant network, or grid. The grid connects most homes to power plants that generate electricity from nonrenewable fossil fuels, like coal and oil, and renewable energy sources, like wind and water. Homes connected to the grid receive a steady supply of electrical energy. Thanks to that electricity, the water in your bath is warm, and the food in your refrigerator is cool. And just a flick of a switch lights up a lamp or turns on the TV.

What about homes not connected to the grid?

America has thousands of them isolated high in the mountains or in remote deserts. It is expensive to connect these homes to power lines. Stringing up the wires that carry electricity can cost thousands of dollars for every kilometer wired. Instead, many people live "off the grid." Does that mean they never read at night, watch television, take hot baths, or drink cold milk? No way! In places like the Navajo Nation, people make their own electricity. There in the sunny Southwest, many homes use solar panels to produce electrical energy. Batteries store any extra electricity for cloudy days and dark nights. No grid? No problem!

THE EXPERT

Stan Atcitty

Sandia National Laboratories

Electrical Engineer

▲ **From the time he said his first words as a toddler until he started kindergarten, Stan spoke Navajo at home.**

Stan Atcitty grew up in the Navajo Nation, in a remote and beautiful area near Shiprock, New Mexico. His family didn't have much money, so his mom sometimes wove and sold traditional rugs for grocery money. Stan made things, too, including some of his own toys.

Stan thought he would always stay on the reservation, but his family encouraged him to consider other paths in life. So Stan started community college, where classes about electricity became his favorites. "You couldn't see it, but you could see its output—movement, sound, light," he says. Stan later transferred to a university. He didn't think he could match the other students—until they asked *him* for homework help. "And I thought I was just a reservation kid who didn't know anything about science!" Stan says.

After years of study, Stan is now an expert in power systems used in remote areas. These systems take electricity generated by solar panels or stored in batteries and convert it for use. He's also interested in using energy wisely and efficiently, and sharing the message. "I tell my kids, 'Leave the lights on, and that's less money available to us!'"

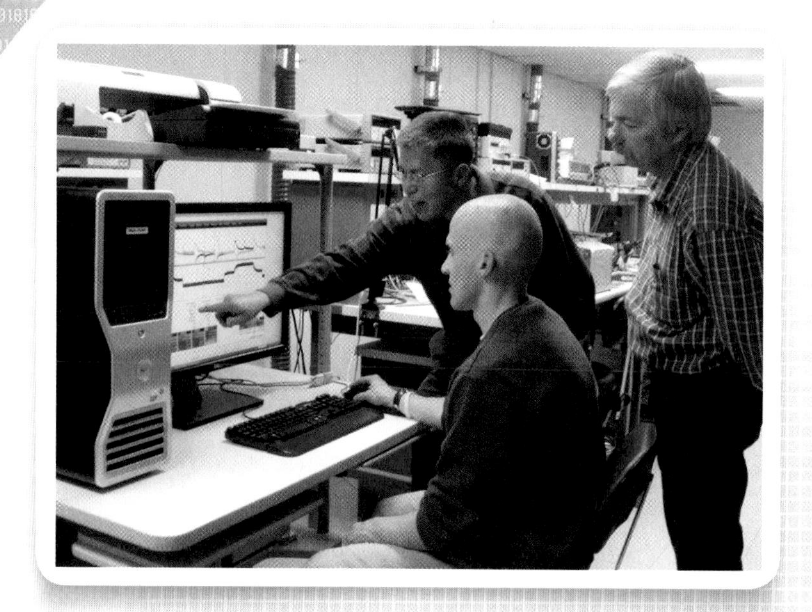

Thanks to computer models, Stan and his team can tinker with a home's energy supply and demand.

The house goes dark. "*Uh-oh.* Out of electricity!" Stan Atcitty says. Everything shuts off—not just the lights, but the refrigerator, television, computer, microwave, and washing machine, too. The house can't get electricity from the grid—the nearest power lines are miles away. The solar panels can't generate more electricity until the Sun comes up. Worse, the backup batteries are drained.

Luckily, this is a simulation. Stan uses a computer model to test the reliability of off-grid power systems. The simulation shows Stan the importance of using energy wisely.

So he shuts off some virtual appliances. "The laundry and microwave popcorn can wait," he says. Now will the batteries last through the night?

Navajos who live in remote areas can make their own energy from wind and Sun.

TECHNOLOGY

Nearly all power plants generate electricity by spinning turbines. Gusting wind, rushing water, or billowing steam gets the turbines spinning. Once the spinning starts, the turbine turns a shaft wound with metal wire surrounded by a magnet. The wire turning inside the magnetic field generates an electric current. *Zoom!* Electricity flows from the generator to the grid.

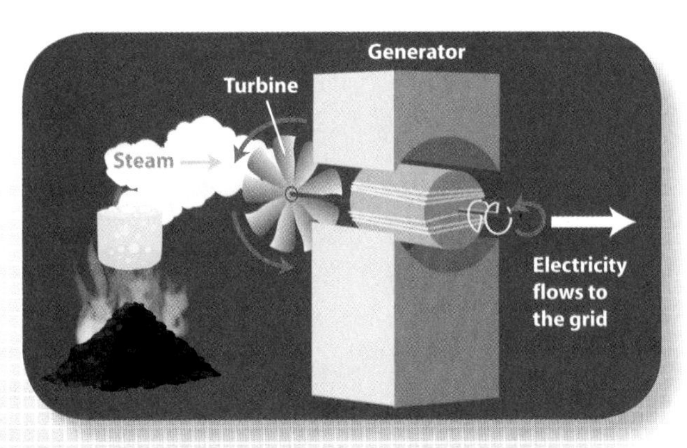

▲ Some power plants produce steam by heating water with a renewable energy source like sunlight. But most power plants produce steam by burning coal or other nonrenewable energy sources.

MATH CONNECTION

Use It Wisely

Electrical power is measured using a unit called the watt. The energy provided by electricity is measured using the watt-hour. It equals 1 watt of power used for 1 hour. An off-grid power system might generate just 3,000 watt-hours a day. How would you use them? Copy the chart and divide up your 3,000 watt-hours.

Appliance	Watts	Hours/Day	Watt-hours
Computer	100	?	?
Television	150	?	?
Refrigerator	150	10	1,500
Washing machine	500	?	?
Coffeemaker	1,000	?	?
Microwave	1,500	0.08	120
Reading lamp	100	?	?
Total			3,000

Oh, By the Way

When Stan was growing up, his family got their drinking water from a barrel filled at a distant well. The more they used, the sooner someone would have to fill the barrel. Stan likes to use this story when explaining how to conserve electricity from an off-grid power system. "You have to be careful how you use up the barrel—the 'barrel' of electricity!" Stan says.

Hey, I Know That!

You've learned a lot about energy—what it is, how it changes, and how it affects your life. On a sheet of paper, show what you know as you do the activities and answer these questions.

1. Look at the diagram. At what point would the skateboarder have the most stored energy? Explain. (page 9)
2. At what point would the skateboarder have the most energy of motion? Explain. (page 9)
3. At what point would the skateboarder's kinetic energy start turning back into stored energy? Explain. (page 9)
4. Describe a series of actions you did or saw happen that involved energy changing forms at least three times. Describe how the energy changed. (pages 10–16)
5. What happens to the energy of a moving bicycle as it rolls along a gravel path? (page 17)
6. How does the Sun produce wind? (page 21)
7. Look back at the pictures on pages 4 and 5. Choose three things in the pictures and describe the kind of energy each has or is using.

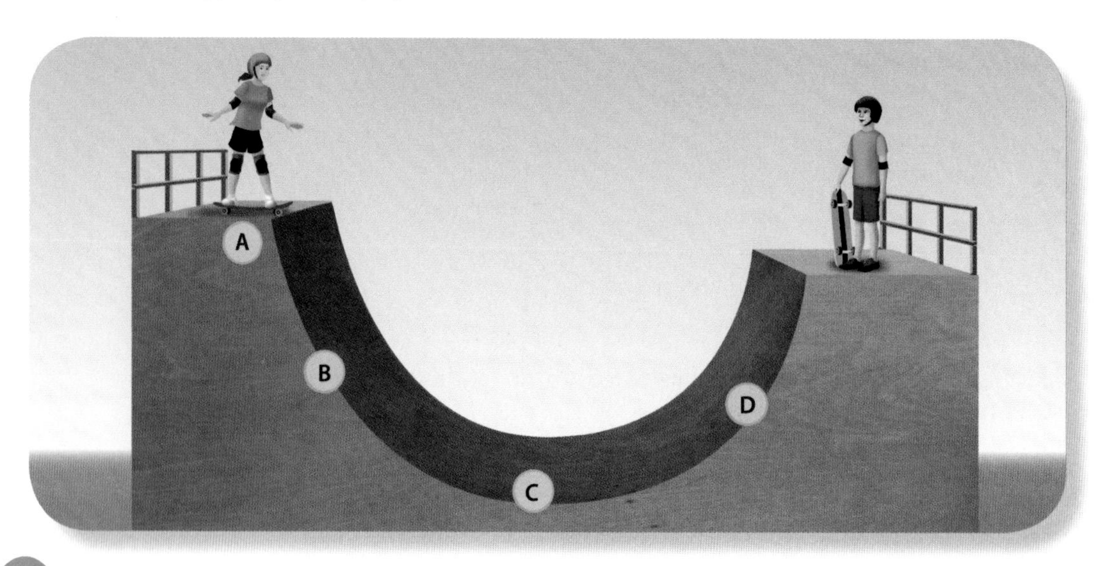

Glossary

chemical energy **(n.)** a form of stored energy released during chemical reactions (p. 12)

chemical reaction **(n.)** a process where one or more substances are changed into new substances with different properties (p. 12)

condense **(v.)** to change gradually from a gas to a liquid (p. 22)

electrical energy **(n.)** energy that comes from the flow of charged particles (p. 15)

energy **(n.)** the ability to cause change and make something happen (p. 5)

evaporate **(v.)** to change gradually from a liquid to a gas (p. 22)

greenhouse effect **(n.)** the trapping of heat by certain gases such as carbon dioxide in Earth's atmosphere. This process warms the air, land, and oceans. (p. 20)

heat **(n.)** energy that comes from the motion of particles that make up matter. Heat flows from an object of higher temperature to an object of lower temperature. (p. 12)

kinetic energy **(n.)** energy that something has because of its motion (p. 10)

law of conservation of energy **(n.)** the principle that energy cannot be created or destroyed—it can only change form and location (p. 17)

light **(n.)** the form of energy associated with the kinds of light our eyes can see. Red light, blue light, and the rest of the colors we can see are known as visible light. Visible light is one part of sunlight. (p. 12)

mass **(n.)** a measure of the total amount of matter contained within an object (p. 11)

photosynthesis **(n.)** the process by which plants and other photosynthetic organisms use energy from sunlight to build sugar from carbon dioxide and water. As part of this process, oxygen is released. (p. 19)

radiate **(v.)** to send out waves of heat or light (p. 20)

solar energy **(n.)** the energy radiated by the Sun in the form of light and heat that is used by plants, phytoplankton, and some bacteria in photosynthesis and harnessed by people to produce electricity (p. 20)

sound **(n.)** longitudinal waves that can be heard by the human ear. Sound waves need to travel through a medium, such as air or water. (p. 12)

turbine **(n.)** a hub with blades that can be rotated by fast-moving air, water, or steam and used to generate electricity (p. 24)

vibrate **(v.)** to move with a quick back-and-forth motion (p. 14)

water cycle **(n.)** the constant movement of water molecules between the air; land; oceans, rivers, and lakes; and living things (p. 22)

watt **(n.)** the main unit used to measure electrical power. One watt is equal to 1 joule of work done over a period of 1 second. (p. 15)

wave **(n.)** a disturbance traveling through a medium like air or water by which energy is transferred from one place to another without moving the medium itself (p. 9)

Index

About the Author Glen Phelan's fascination with science was sparked when he was a teenager by the lunar missions of the Apollo Program. He shares his fascination through teaching and writing. Learn more at www.sallyridescience.com.

Photo Credits TebNad/Shutterstock.com: Cover. ilker canikligil/Shutterstock.com: Back cover. Reuters/Ho New: Title page, p. 10. Wiktor Bubniak/Shutterstock.com: p. 2. karnizz/Shutterstock.com: p. 4. mountainpix/Shutterstock.com: p. 5. Andraž Cerar/Shutterstock.com: p. 6. Aaron Amat/Shutterstock.com: p. 7 top. hfng/Shutterstock.com: p. 7 bottom. Mihajlo Maricic/iStockphoto.com: p. 9 top. Robert Bremec/iStockphoto.com: p. 9 bottom. Ocean Photography: p. 11. NASA/Carla Cioffi: p. 12. XAOC/Shutterstock.com: p. 13 (gasoline). Sally Wallis/Shutterstock.com: p. 13 (coal). grzym/Shutterstock.com: p. 13 (wood). CoolR/Shutterstock.com: p. 13 (car). Krystian Konopka/Shutterstock.com: p. 13 (power lines). LianeM/Shutterstock.com: p. 13 (fire). Etienne du Preez/Shutterstock.com: p. 15 top. Nikada/iStockphoto.com: p. 15 bottom, p. 17. Doug Lemke/Shutterstock.com: p. 16. Triff/Shutterstock.com: p. 18 (sky). Dee Breger/Science Photo Library: p. 18 (phytoplankton). Pete Bucktrout/www.photo.antarctica.ac.uk: p. 18 (krill). idreamphoto/Shutterstock.com: p. 18 (whale). NASA Goddard Space Flight Center: p. 20 (globe in illustration). Tania Zbrodko/Shutterstock.com: p. 21 top. NASA: p. 21 bottom. Norio Yamagata/Nature Production/Minden Pictures: p. 22. Central Intelligence Agency: p. 25 (map in illustration). Mike Norton/Shutterstock.com: p. 26. Courtesy of Stan Atcitty: p. 27, p. 28 top. Courtesy of Sandia National Laboratories. Sandia National Laboratories is a multiprogram laboratory managed and operated by Sandia Corporation, a wholly owned subsidiary of Lockheed Martin Corporation, for the U.S. Department of Energy's National Nuclear Security Administration under contract DE-AC04-94AL85000.: p. 27 (logo). NTUA Hybrid Solar Power System on the Navajo Nation: p. 28 bottom.